ON THE VARIETIES OF MAN

IN THE MALAY ARCHIPELAGO

(1864)

BY

ALFRED RUSSEL WALLACE

British Library Cataloguing-in-Publication Data
A catalogue record for this book is available from the
British Library

Alfred Russel Wallace

Alfred Russel Wallace was born on 8th January 1823 in the village of Llanbadoc, in Monmouthshire, Wales.

At the age of five, Wallace's family moved to Hertford where he later enrolled at Hertford Grammar School. He was educated there until financial difficulties forced his family to withdraw him in 1836. He then boarded with his older brother John before becoming an apprentice to his eldest brother, William, a surveyor. He worked for William for six years until the business declined due to difficult economic conditions.

After a brief period of unemployment, he was hired as a master at the Collegiate School in Leicester to teach drawing, map-making, and surveying. During this time he met the entomologist Henry Bates who inspired Wallace to begin collecting insects. He and bates continued exchanging letters after Wallace left teaching to pursue his surveying career. They corresponded on prominent works of the time such as Charles Darwin's *The Voyage of the Beagle* (1839) and Robert Chamber's *Vestiges of the Natural History of Creation* (1844).

Wallace was inspired by the travelling naturalists of the day and decided to begin his exploration career collecting specimens in the Amazon rainforest. He explored the Rio Negra for four years, making notes on the peoples and languages he encountered as well as the geography, flora, and fauna. On his return voyage his ship, Helen, caught fire and he and the crew were stranded for ten days before being picked up by the Jordeson, a brig travelling from Cuba to London. All of his specimens aboard Helen had been lost.

After a brief stay in England he embarked on a journey to the Malay Archipelago (now Singapore, Malaysia, and Indonesia). During this eight year period he collected more than 126,000 specimens, several thousand of which represented new species to science. While travelling, Wallace refined his thoughts about evolution and in 1858 he outlined his theory of natural selection in an article he sent to Charles Darwin. This was published in the same year along with Darwin's own theory. Wallace eventually published an account of his travels *The Malay Archipelago* in 1869, and it became one of the most popular books of scientific exploration in the 19th century.

Upon his return to England, in 1862, Wallace became a staunch defender of Darwin's landmark work *On the Origin of Species* (1859). He wrote responses to those critical of the theory of natural selection, including 'Remarks on the Rev. S. Haughton's Paper on the Bee's Cell, And on the Origin of Species' (1863) and 'Creation by Law' (1867). The former of these was particularly pleasing to Darwin. Wallace also published important papers such as 'The Origin of Human Races and the Antiquity of Man Deduced from the Theory of 'Natural Selection" (1864) and books, including the much cited *Darwinism* (1889).

Wallace made a huge contribution to the natural sciences and he will continue to be remembered as one of the key figures in the development of evolutionary theory.

Wallace died on 7th November 1913 at the age of 90. He is buried in a small cemetery at Broadstone, Dorset, England.

ON THE VARIETIES OF MAN IN THE MALAY ARCHIPELAGO

(1864)

Stretching out from the south-eastern parts of Asia, and extending over two-thirds of the great Pacific Ocean, is a region of innumerable islands, the inhabitants of which, though notoriously differing widely among themselves, have been considered by almost all ethnologists to form one of the strongly-marked varieties into which the human race can be divided.

The term Oceania has been aptly applied to this portion of the globe, and its native tribes have been collectively classed as the Malayan or Malayo-Polynesian Races.

In the present paper I propose to give some account, from personal observation, of the inhabitants of the chief islands of the Malay Archipelago, and by a comparison with the published descriptions of the inhabitants of the surrounding countries, to endeavour to arrive at some definite conclusion as to their mutual relations or their common origin.

Before entering upon the description of these races of

man, I must say a few words about the country they inhabit, and point out some of the most interesting and important peculiarities in its forms of animal and vegetable life.

This Malayan region is indeed remarkable in many respects. It is the largest Archipelago in the world. It contains the two largest islands in the world, one of which, Borneo, could embrace within its limits the whole of the British Isles from the Land's End to the Orkneys, and surround them on every side with a green ocean of tropical forest. It contains, in the great volcanic belt that runs through its whole extent, a vast number of active volcanoes, and is unequalled for the frequency of its eruptions and earthquakes; while one island, Java, may claim to be superlative in almost everything, since it has more volcanoes than any other tract of country of equal extent upon the globe, its vegetation is most luxuriant and varied, its animal productions most abundant and beautiful, and its native inhabitants are more numerous and of a better disposition than those of perhaps any other tropical country. It is certainly one of the most fertile, and I believe it to be one of the best governed of all tropical lands under European sway.

Throughout the whole Archipelago, beauty of vegetation is a pre-eminent charm. Almost every island, as well as every mountain peak, is clothed with the most luxuriant tropical forest, amid which, palms and tree ferns and the broad-leaved Musaceæ (the most noble and elegant of vegetable forms) are always a conspicuous feature; while every native village is embosomed in groves of fruit trees which attain almost the same altitude and luxuriance as the virgin forests. Among the individual products of the islands most worthy of note, are the precious spices, nutmeg, cloves, and cinnamon, the sago palm, the rattan cane, the fragrant sandal wood, and the gutta percha trees; while the delicious mangusteen, the queen of fruits, cannot even be successfully cultivated beyond these favoured regions.

In the animal world, the most remarkable productions are the man-like orang-utang, found only in Borneo, and the lovely birds of Paradise, confined to the remote islands of New Guinea; while edible birds' nests and mother-of-pearl-shell are valuable and interesting products almost restricted to this region. It was my good fortune to spend eight years in these beautiful islands. I visited about thirty of them and nearly one hundred towns and villages, sailing in native praus among their interesting but dangerous coral reefs, roaming through their ever-verdant forests, and dwelling in the rude huts of their uncivilised inhabitants. My journeys from island to island within the Archipelago amounted to a total of about fifteen thousand miles (or considerably more than half the circumference of the globe), of which more than four thousand miles were made in open boats and native praus, constructed without the use of iron and fastened together with wooden pegs and rattans.

The most important physical feature in this Archipelago, and the one that has had most effect in determining the character of its natural productions, is hidden beneath the waters; but as it has probably had some influence in determining the distribution of mankind (as it certainly has of the animal races), I must briefly allude to it. From Asia on the one side, and from Australia on the other, stretch out immense submarine banks, which connect the adjacent islands to the main land; while beyond the limits of these banks an unfathomable ocean is always found. The great islands of Sumatra, Java, and Borneo, are situated on the Asiatic banks; and from the fact that a number of the largest animals, which could not have crossed the sea (as the rhinoceros of Java and Sumatra, the elephant of Sumatra, the tapir of Sumatra and Borneo, and the wild ox of Java) being all found on some part or other of the Asiatic Continent, it is rendered highly probable, if not certain, that all these great islands have very recently formed a part of Asia, from which they have been separated by the same convulsions which have raised up the great chains of volcanic mountains in Sumatra and Java. In the same manner, and for similar reasons, New Guinea and some adjacent islands have been once connected with Australia.

The belt of deep ocean, however, which intervenes between these banks indicates separation for very long periods, and there is evidence in the geological structure of

Java and Borneo to show that some parts of these islands are of recent formation, so that the gulf between the Asiatic and Australian regions was formerly much greater than it is now. This is most strikingly confirmed by the almost total diversity between the animals of the two halves of the Archipelago. In the western islands all the chief forms of animal life of the Asiatic continent are found, while in the eastern half, we have chiefly those characteristic of Australia; so that a line drawn across the Archipelago west of Celebes marks the division between two great zoological regions of the globe.

The amount of diversity between these two regions is fully as great as exists between Africa and South America; and this diversity is to a great extent maintained even between those border lands which approach within sight of each other. The islands of Baly and Lombock are only fifteen miles apart, yet they really belong to two distinct divisions of the globe. The naturalist who is accustomed to distinguish characteristic forms of animal life, is at once struck with the remarkable difference between them. In Baly he is still in the Indian region of the Archipelago; the common birds which he sees are *barbets*, black and white thrushes, crested starlings, fruit thrushes, and woodpeckers,--all of which are common forms in India, Sumatra, Borneo, and Java. He may see all these in the morning, and in the afternoon cross over to Lombock where a totally different set of birds will present themselves. He will now see *white cockatoos, long-necked honey-suckers,* and the *mound-making megapodius,*--all common forms in the islands farther east and in Australia, but not one of which ever passes across this narrow strait to the islands further west. If larger and more distant islands are compared, the difference is more complete in every department of nature,--so that between Borneo and New Guinea, though almost exactly agreeing in climate and physical features, there is a total diversity of animal productions--a diversity as complete and striking as that which exists between the old and new worlds.[1]

Having thus briefly indicated the most important features in the animal productions of these regions, we will now turn to the consideration of man.

Two very strongly contrasted races are found in this region: the *Malays* who inhabit almost exclusively the western half of the Archipelago, and the *Papuans* whose head-quarters are New Guinea and some of the adjacent islands. In many of the remaining islands the inhabitants differ considerably from both of these primary races, but can generally be referred as modifications of one or the other; we will, therefore, commence by pointing out those characteristic peculiarities which distinguish the Malayan and Papuan races.

Malays. The colour of the typical Malay is a light reddish-brown, with a more or less olive tint. The colour may be compared to that of cinnamon or half-roasted coffee, and it is on the whole very constant over a wide extent of country. The inhabitants of the Malay peninsula, of Sumatra, of Java, of Celebes, and of the Philippines, do not differ more in colour from each other than do the people of the different countries of Europe. The hair is also very constant, being invariably black and straight, so much so that any lighter tint, or any wave or curl however slight, is rarely met with, and may I believe be considered a sure proof of the admixture of some foreign blood. The face is almost always destitute of beard; sometimes a small one is produced in middle or old age, but as a general rule in young men it is almost entirely absent. Among some of the tribes it is cultivated with great care, and the very small quantity that can then be produced is the strongest proof of its natural deficiency. I have seen a native of Java who had evidently taken great pains to develope moustaches and beard, but the component hairs of the former were so few that when well waxed up, they formed but a slender thread several inches long, and the beard was still more scanty, having only been produced from a mole on one side of the chin. Yet this man was very proud of his beard and moustaches, which he was continually fingering, as if to call attention to such an unusual and remarkable adornment. The arms, legs, and breasts of this people are, to use a common expression, as smooth as the palm of one's hand, and almost as completely destitute of any growth of hair.

The stature of the Malay is generally a few inches below that of the middle-sized European: the body is robust; the breast much developed; the feet small, thick, and short; the hands small and rather delicate. The face is a little broad and inclined to be flat; the forehead is rather rounded, the brows low, the eyes black and very slightly oblique; the nose is rather small, not prominent but straight and well-shaped, the apex a little rounded, the nostrils broad and slightly exposed; the cheek bones are rather prominent, the mouth large, the lips broad and well cut but not protruding, the chin round and well-formed.

In this description there seems little to object to on the score of beauty, and yet on the whole the Malays are certainly not handsome. In youth, however, they are often very good-looking, and many of the boys and girls up to twelve or fifteen years of age are very pleasing, and some have countenances which are in their way almost perfect. I am inclined to think they lose much of their good looks by bad habits and irregular living. At a very early age they chew betel and tobacco almost incessantly; they suffer much want and exposure in their fishing and other excursions; their lives are often passed in alternate starvation and feasting, idleness and excessive labour,--and this naturally produces premature old age and harshness of features.

In character the Malay is impassive. He exhibits a reserve, diffidence, and even bashfulness, which is in some degree attractive, and leads the observer to think that the ferocious and blood-thirsty character imputed to the race must be grossly exaggerated. He is not demonstrative. His feelings of surprise, admiration, or fear, are never openly manifested, and are probably not strongly felt. He is slow and deliberate in speech, and circuitous in introducing the subject he has come expressly to discuss. These are the main features of his moral nature, and exhibit themselves in every action of his life.

Children and women are timid, and scream and run at the unexpected sight of a European. In the company of men they are silent, and are generally quiet and obedient. When alone the Malay is taciturn; he neither talks nor sings to himself. When several are paddling in a canoe, they occasionally chant a monotonous and plaintive song. He is cautious of giving offence to his equals. He does not quarrel easily about money matters; dislikes asking too frequently even for payment of his just debts, and will often give them up altogether rather than quarrel with his debtor. Practical joking is utterly repugnant to his disposition; for he is particularly sensitive to breaches of etiquette or any interference with the personal liberty of himself or another. As an example, I may mention that I have often found it very difficult to get one Malay servant to waken another. He will call as loud as he can, but will hardly touch much less shake his comrade. I have frequently had to waken a hard sleeper myself when on a land or sea journey. Even when a criminal is brought for trial, it is amusing to see how tenderly he is often treated by his comrades who are acting as policemen. He is nominally bound, but a short piece of cord is merely once turned round his hands, which he can shake off whenever he pleases. While the examination is going on, he sits down on the mat along with witnesses and police, and as the punishment is generally a fine, the whole matter passes off with the least possible interference with the prisoner's personal liberty.

The Malay seems to have little appreciation of the ludicrous, and does not laugh heartily. Little accidents to his comrades excite his merriment most, and in listening to a tale, the bold attack or treacherous massacre seem to him the most amusing incidents. He never laughs and rarely even smiles at Europeans, but however strange their actions may seem to him, he expresses it only by a stare of surprise or contempt. He rarely expresses joy or gratitude openly. If you make him a present, he receives it in silence; and if, through ignorance of the state of the market, you pay him twice as much as he expects to receive, you will never find out your mistake by an examination of his countenance. He will revenge an insult sooner than he will resent an injury, and will do so by secret treachery rather than by an open attack. He is honest if trusted with property, but looks upon bold and skilful lying as rather meritorious than otherwise.

The Malay is not much affected by the beauties of external nature and has no sense of order. In his villages the houses are scattered about irregularly, and the planted fruit trees grow crowded in the most inconvenient places. He loves sweet-smelling flowers, but never cultivates a plant for beauty alone. He looks upon a mountain as very inconvenient to climb, and upon a foaming cataract as a great impediment to navigation, but cannot conceive that there is anything to admire in the one or the other.

In intellect the Malay is but mediocre. He is deficient in the energy requisite for the acquisition of knowledge, and seems incapable of following out any but the simplest combinations of ideas. If you give him instructions involving different courses of action dependent on contingent events, he is almost sure to misunderstand them; and anything beyond the simplest calculations cannot be made comprehensible to him. He is capable of considerable proficiency in mechanical arts, and therefore makes a good servant for performing any established routine of duty, but can seldom be trusted in matters in which judgment or discretion are necessary.

The state of civilisation reached by the Malay races is not very high. A considerable number of tribes in Borneo, Sumatra, and Malacca, are what we call savages. They wear the scantiest clothing, often of bark; their only manufactures are a few weapons, canoes, and rude huts; but all are more or less settled and practise agriculture, so that they are not at the lowest stage of barbarism. Some of these tribes weave cloth, others make good native iron, from which they manufacture such excellent weapons that a dyak knife will cut in two an ordinary ship's cutlass. None of these wild tribes have any government other than that of village chiefs who have only a nominal authority, and it is this that really constitutes them savages. The larger half of the Malay races have, however, risen above this condition, and have rajahs or sultans ruling over more or less extensive territories, with princes and chiefs beneath them with something like established laws and courts of justice. The people in this condition all have a written language and a rude literature consisting of legends, songs, and dramatic poems. The true Malays use the Arabic character, while the natives of Celebes, Java, and some parts of Sumatra have each a peculiar indigenous alphabet. It is to be remarked, however, that all these more civilised peoples are Mahometans or of the Brahminical religion, and it therefore seems highly probable that the civilisation they have attained is not altogether spontaneous.

Papuans. The typical Papuan race is in many respects the very opposite of the Malay, and it has hitherto been very imperfectly described. The colour of the Papuan is a deep sooty-brown or black, sometimes approaching but never quite equalling the jet-black of some Negro races. It varies in tint, however, more than that of the Malay, and is sometimes a dusky-brown. The hair is very peculiar, being harsh, dry, and frizzly, growing in little tufts or curls, which in youth are very short and compact, but afterwards grow out to a considerable length, forming the compact frizzled mop which is the Papuans' pride and glory. The face is adorned with a beard of the same frizzly nature as the hair of the head. The arms, legs, and breast are also more or less clothed with hair of a similar nature.

In stature the Papuan decidedly surpasses the Malay, and is perhaps equal to the average of Europeans. The legs are long and thin, and the hands and feet larger than in the Malays. The face is somewhat elongated, the forehead flattish, the brows very prominent; the nose is large, rather arched and high, the base thick, the nostrils broad with the aperture hidden, owing to the tip of the nose being elongated; the mouth is large, the lips thick and protuberant. The face has thus an altogether more European aspect than in the Malay, owing to the large nose; and the peculiar form of this organ, with the more prominent brows and the character of the hair on the head, face, and body, enable us at a glance to distinguish the two races. I have observed that most of these characteristic features are as distinctly visible in children of ten or twelve years old as in adults.

The moral characteristics of the Papuan appear to me to separate him as distinctly from the Malay as do his form and features. He is impulsive and demonstrative in speech and action. His emotions and passions express themselves in shouts and laughter, in yells and frantic leapings. Women and children take their share in every discussion, and seem little alarmed at the sight of strangers and Europeans. My first acquaintance with this race was made at the Ké Islands, and it was a scene I shall never forget. As we approached the shore (in a Malay prau from Macassar) three or four native boats came out to meet us. The crews were singing and paddling most furiously, and as they neared us they increased their shouts and danced and made strange gestures. In a few moments more they were alongside and all scrambled up on deck, and then commenced a scene of indescribable confusion. They yelled and screamed and jabbered, danced and leaped and ran about as if they were out of their senses with joy and excitement. They rushed from one sailor to another begging tobacco and rice, and a score of them surrounded our captain and almost pulled him to pieces, offering to tow his vessel into the harbour and begging immediate payment, all talking at once and each one struggling for the first place. Between whiles they would talk to each other with violent gestures, laugh and grin with delight when they got anything, roll on the deck and jump up again, or leap headlong overboard and be upon our vessel again in an instant, as if merely to work off some of the superabundant energy and excitement with which they were charged.

To myself, who had lived several years among Malay races,

this conduct was very striking. Under similar circumstances Malays would have come up in perfect silence. On deck they would have stood still or sat down with a simple salutation, and would have waited patiently for us to address any inquiries to them we thought proper.

A few days afterwards, when on shore at Ké, I met an old man some distance inland. He saw me catching a butterfly and stopped to see what I was doing. He watched carefully and with a puzzled expression, while I got the insect out of my net, took a pin from my pincushion, pinned it, and secured it in my collecting-box, when he could contain himself no longer, but bending his body almost double enjoyed a hearty roar of laughter. Scores of times have I been met by Malays under similar circumstances and to whom the work I was engaged in was equally novel, but not one of them ever shewed the slightest inclination to laugh, but contented themselves either with asking gravely what it was for, or passing on with a quiet stare.

The constant activity and noisiness of the Papuans would alone shew that they are a distinct race from the Malays. I lived for two months in a native house in the interior of the Aru Islands. It was occupied by about five families and from morning till late at night there was a continual row. My Malay servants were quite astonished, and used to remark to me "What strong talkers these Aru men are!" And in fact, I was lately reminded of my residence here by

Dr. Livingstone's account of the incessant worry and noise kept up by the natives of South Africa. Of the intellect of these people it is very difficult to judge, but I am inclined to rate it higher than that of the Malays, notwithstanding the fact that the Papuans have never yet made any advance towards civilisation. It must be remembered, however, that for centuries the Malays have been influenced by Hindoo, Chinese, and Arabic immigration, whereas the Papuan race has only been subjected to the very partial and local influence of Malay traders. The Papuan has much more vital energy, which would certainly greatly assist his intellectual development. Papuan slaves show no inferiority of intellect compared with Malays, but rather the contrary; and in the Moluccas they are often promoted to places of considerable trust. The Papuan has a greater feeling for art than the Malay. He decorates his canoe, his house, and almost every domestic utensil with elaborate carving, a habit which is rarely found among tribes of the Malay race.

In the affections and moral sentiments, on the other hand, the Papuans seem very deficient. In the treatment of their children they are often violent and cruel; whereas the Malays are almost invariably kind and gentle, hardly ever interfering at all with their children's pursuits and amusements, and giving them perfect liberty at whatever age they wish to claim it. But these very peaceful relations between parents and children are no doubt, in a great measure, due to the listless and apathetic character of the race, which never leads the younger members into serious opposition to the elders; while the harsher discipline of the Papuans may be chiefly due to that greater vigour and energy of mind which always sooner or later leads to the rebellion of the weaker against the stronger, the people against their rulers, the slave against his master, or the child against his parent.

It appears therefore, that whether we consider their physical conformation, their moral characteristics, or their intellectual capacities, the Malay and Papuan races offer remarkable differences and striking contrasts. The Malay is of short stature, brown skinned, straight haired, beardless, and smooth bodied; the Papuan is taller, is black skinned, frizzly haired, bearded, and hairy bodied; the former is broad faced, has a small nose and flat eyebrows; the latter is long faced, has a large and prominent nose, and projecting eyebrows. The Malay is bashful, cold, undemonstrative, and quiet; the Papuan is bold, impetuous, excitable, and noisy; the former is grave and seldom laughs; the latter is joyous and laughter-loving,--the one conceals his emotions, the other displays them.

Where shall we find races offering more remarkable contrasts than these? If mankind can be classed at all into distinct varieties, surely the Malays and Papuans must be kept for ever separate.

I will now give some account of the inhabitants of the

numerous islands which do not exactly agree with either of these two great races.

The typical Malays, as has been before mentioned, occupy the whole of the Malay peninsula, Sumatra, Borneo, the Philippine Islands, Java, Baly, Lombock, Sumbawa, and Celebes. These islanders exhibit slight differences among themselves, so that an experienced observer can generally point out a Dyak, a Bugis, or an inhabitant of the Philippines, but the distinguishing characters are evanescent, and it is almost impossible so to fix upon them as to render them capable of description.

Beyond Celebes the islands which have a Malayan population are the Sulla Isles, Batchian, and the small islands up to Tidore and Ternate. The Sulla Isles are inhabited by a race of Aborigines resembling those of the nearest peninsula of Celebes. These people are short, broad-faced, industrious, and mild in their disposition. Some of them have been recently brought by the Dutch Government to Batchian, to save them from the extermination with which they were threatened by their more warlike neighbours in the interior.

Before leaving the Malayan region, I must mention the northern peninsula of Celebes, of which the chief town is Menado, and which is inhabited by a race quite distinct from the other people of the island. In many respects this little district is one of the most interesting in the whole archipelago. It is volcanic, and possesses a very rich and fertile soil, whose fertility is increased by the abundant moisture which results from its situation on the equator. The whole interior is a plateau of from 2,000 to 3,000 feet elevation, just sufficient to temper the tropical heat and produce a perpetual summer. And this favoured district is inhabited by a people who are of a mild disposition, of a pleasant countenance; who are tractable, industrious, and intelligent. In all the surrounding islands one hears tell of the people of Minahassa, who are said (with a little exaggeration) to be as white as Europeans, and as good-looking. Thirty years ago these people were savages, exactly in the same condition as the Dyaks of the interior of Borneo. They went naked, with only the same rude dress as the Dyaks; they were continually at war one village with another; they held riotous festivals and practised obscene ceremonies. Their houses were raised high up on tall pillars, to protect them from night attacks, and they were subject to the alternate plenty and famine of the savage state.

The Dutch government, however, took them in hand, settled missionaries among them, and introduced the cultivation of coffee through the intervention of the native chiefs, whose power was upheld by giving them the title of "major," and whose influence in favour of the plan was secured by a per-centage on the proceeds. One sees the result now in a beautiful and well cultivated country, in neat and regular villages, and good roads; in a population well fed and well clothed, the greater part of whom are Protestant Christians, who can most of them read and write, and who, if they please, can enjoy a great many of the comforts and luxuries of civilisation.

These people are Malays, but have something of the Tartar and something of the European in their physiognomy. They agree best with some of the inhabitants of the Philippines, and I have no doubt that they have come from those islands originally by way of the Siaou and Sanguir Islands, which are inhabited by an allied race. Their languages show this affinity, differing very much from all those of the rest of Celebes. A proof, however, of the antiquity of this immigration, and of the low state of civilisation in which they must have existed for long periods, is to be seen in the variety of their languages. In a district about one hundred miles long by thirty miles wide, not less than ten distinct languages are spoken. Some of these are confined to single villages, others to groups of three or four; and though of course they bear a certain family resemblance, they are yet so distinct as to be mutually unintelligible.

Here I think, we have a proof that the absence of civilisation does not necessarily imply the want of capacity to receive it. An external impulse is in every case required; for I believe no instance can be shown of a homogeneous race having made much or any progress when uninfluenced by the contact of other races. Civilisation has ever accompanied emigration and conquest--the conflict of opinion, of religion, or of race. In proportion to the diversity of these mingling streams, have nations progressed in literature, in arts, and in science; while, on the other hand, when a people have been long isolated from surrounding races, and prevented from acquiring those new ideas which contact with them would induce, all progress has been arrested, and generation has succeeded generation with almost the same uniformity of habits and monotony of ideas as obtains in the animal world, where we impute it to that imaginary power which we designate by the term *instinct*.

The true Moluccas, comprehending the islands of Ternate, Tidore, Makian, and Batchian, are inhabited by people who possess the chief Malayan characteristics, and who may be considered to mark the extreme eastern range of the true Malay race. These people are all Mahometans; they have the usual Malay civilisation and a regular form of government. There are three independent chiefs, with the title of Sultans of Batchian, Tidore, and Ternate. The first rules over the island of Batchian and the small islands only that surround it, including Obi. The sultan of Ternate possesses also the southern extremity of Gilolo opposite Batchian, the entire northern peninsula of that island, and the islands of Morty, Makian, and the Sulla Islands. The sultan of Tidore governs the remainder of Gilolo, all the islands between it and New Guinea, and the whole coast of the northern peninsulas of that great island.

The islands of Obi, Batchian, and the three southern peninsulas of Gilolo possess no true indigenous population. The northern peninsula, however, is inhabited by a native race, whose principal tribes are the so-called Alfurus of Sahoe and Galela. These people are quite distinct from the Malays, and almost equally so from the Papuan. They are tall and well-made, with Papuan features, curly hair, bearded and hairy-bodied, but quite as light in colour as the Malays. They are an industrious and enterprising race, cultivating largely, and indefatigable in their search after the natural productions of sea and land. They are very intelligent and improvable, and on the whole seem one of the best races in the archipelago. From the fact of the greater part of Gilolo and many of the smaller islands being still quite without indigenous inhabitants, and from the rude civilisation attained by the inhabitants of the Moluccas at a very early period, I have come to the conclusion that the kingdoms of Tidore and Ternate were founded by some roving Malayan princes, who, being charmed with the bright calm seas and fertile soil of these volcanic islets, settled permanently upon them, and upon the coasts of the adjoining island of Gilolo, over the whole of which they gradually assumed the sovereignty. From their constant intercourse with the indigenous inhabitants of Gilolo they would derive their languages, in which but a small portion of their original Malay would remain, just as the invaders of Normandy acquired the tongue of the people they conquered, and changed their language a second time, when they invaded and settled in Saxon England.

In the great island of Ceram an indigenous race also exists, very similar to that of Northern Gilolo. Bouru seems to contain two races, a shorter rounder-faced people with a Malay physiognomy, who have probably come from East Celebes by way of the Sulla islands, and a taller bearded race resembling that of Ceram.

A singular mixed race must be mentioned as characteristic of the Moluccas, the "Orang Sirani," or native Christians. These have generally separate quarters in the chief towns and villages. They all speak the Malay language, with a considerable admixture of Portuguese; they are tolerable musicians, and are very fond of European dances. These people are no doubt the descendants of the early Portuguese settlers. They retain much of European physiognomy and manners, and consider themselves to be far superior to the native races. In many places along the coast of Ceram and the adjacent islands they have however, intermingled with the indigenes and with the Malay races, which have spread from Ternate, Batchian, etc., so as to produce inextricable confusion of type. The people of the coasts of all these islands between Celebes and New Guinea are constantly changing about. The one fact, however, that I consider indisputable is, that here the Malay race comes in contact with two other races, which are non-Malay--the Alfurus of Gilolo and Ceram, and the Papuans of New Guinea, Waigiou, Mysol, and the Arru and Ké Islands.

Far south of the Moluccas lays the island of Timor, inhabited by tribes much nearer to the true Papuan than those of the Moluccas.

The Timorese of the interior are dusky brown or blackish, with bushy frizzled hair, and the long Papuan nose. They are of medium height, and rather slender figures. The universal dress is a long cloth twisted round the waist, the fringed ends of which hang below the knee. The people are said to be great thieves, and the tribes are always at war with each other, but they are not very courageous or bloodthirsty. The custom of "tabu," called here "pomáli," is very general, fruit trees, houses, crops, and property of all kinds being protected from depredation by this ceremony, the reverence for which is very great: a palm branch stuck across an open door, showing that the house is taboed, is a more effectual guard against robbery than any amount of locks and bars. The houses in Timor are different from those of most other islands; they seem all roof, the thatch overhanging the low walls and reaching to the ground, except where it is cut away for an entrance. In the west end of Timor, on the little island of Semau, the houses more resemble those of the Hottentots, being egg-shaped, very small, and with a door only about three feet high. These are built on the ground, while those of the eastern districts are raised a few feet on posts. In their excitable disposition, loud voices, and fearless demeanour, the Timorese closely resemble the people of New Guinea.

In the islands west of Timor, as far as Flores and Sandalwood Island, a very similar race is found, which also extends eastward to Timor-laut, where the true Papuan race begins to appear. The small islands of Savu and Rotti, however, to the west of Timor, are very remarkable in possessing a different and in some respects peculiar race. These people are very handsome, with good features, resembling in many characteristics the race produced by the mixture of the Hindoo or Arab with the Malay. They are certainly distinct from the Timorese or Papuan races, and must be classed in the western rather than the eastern ethnological division of the archipelago.

The whole of the great island of New Guinea, the Ké, and Aru Islands, with Mysol, Salwatty, and Waigiou, are inhabited almost exclusively by the typical Papuans. I found no trace of any other people inhabiting the interior of New Guinea, but the coast people are in some places mixed with the browner tribes of the Moluccas. The same Papuan race seems to extend over the islands east of New Guinea as far as the Fidjees.

There remain to be noticed the black woolly-haired races of the Philippines and the Malay peninsula, the "Negritos" and the "Semangs." I have never seen these people myself, but from the numerous accurate descriptions that have been published of them, I have had no difficulty in satisfying myself that they have no affinity or resemblance whatever to the Papuans, with which they have been hitherto associated. In most important characters they differ more from the Papuan than they do from the Malay. They are dwarfs in stature, only averaging four feet six inches to four feet eight inches high, or eight inches less than the Malays; whereas the Papuans are decidedly taller than the Malays. The nose is invariably represented as small, flattened, or turned up at the apex, whereas the most universal character of the Papuan race is to have the nose prominent and large, with the apex produced downwards, as it is invariably represented in their own rude idols. The hair of these dwarfish races agrees with that of the Papuans, but so it does with that of the negroes of Africa. The *Negritos* and the *Semangs* agree very closely in physical characteristics with each other and with the Andaman Islanders, while they differ in a most marked manner from every Papuan race.

Having thus briefly sketched the physical and mental characteristics of the chief inhabitants of the archipelago, it will no doubt be expected that I should state what are the conclusions at which I have arrived as to their origin.

The great attention that has recently been given to the problem of the antiquity of man, and the advance made towards a solution of it, has invested the question of the origin of races with a new interest, and has also furnished to the ethnologist the means of avoiding many of the difficulties which formerly embarassed him.

When even the geologist would only grant us a very limited period for the existence of the human race upon the earth--when he, to a certain extent, supported the popular belief that man had originated but a few thousand years ago, no wonder that the ethnologist found it impossible to account for the vast differences observed in mankind by any natural process of change. Not only have we manners and customs which among the less civilised races change but slowly, but we have languages the most diversified and the most incongruous, which we in vain seek to trace back to a common origin. Not only have we absolute contrasts of colour, of hair, of features, of stature, which neither climate nor other external conditions seem to have sensibly affected during the historical period, but we have mental and moral peculiarities equally marked, producing national characters, which we have still less reason to believe have changed or can change, except with extreme slowness.

These insurmountable difficulties have led many ethnologists to adopt the hypothesis that man is not one but many; that whenever he originated, it was in several localities and under various forms; that, in fact, the chief races of man are aboriginally distinct, and were created as they now are and where they are now found.

Accepting however most gratefully, the permission we now have to place the origin of man at an indefinitely remote epoch, our difficulties are in a great measure removed, and we can speculate more freely on the parentage of tribes and races. We are further enabled to introduce a new element of the greatest importance into our reasonings on this subject--the geological changes of the earth's surface; for, as it is now certainly proved that man coexisted with extinct quadrupeds, and has survived elevations and depressions of the earth's surface to the amount of at least several hundred feet, we may consider the effects of the breaking up or re-formation of continents, and the subsidence of islands, on the migrations, the increase, or the extinction of the people who inhabited them.

We have, moreover, a remarkable instance of a physical change in a people, with whose origin we are well acquainted, going on under our own eyes, and dependent on material and moral causes which we can in some degree trace out. I allude to the peculiar characteristics of the people of the United States of North America, which are sufficiently palpable to be noticed by every writer who has visited them, and are the more extraordinary on account of the variety of races which have contributed to populate the country. We cannot, therefore, deny that man is to some extent changeable, even in short periods of time; and it is very difficult to limit the effects of analogous causes acting through those vast epochs which have been require to bring about the last great changes in the condition of earth's surface.

From considerations such as these, taken in connexion with the physical and moral peculiarities of the races of the archipelago, together with those of Eastern Asia, of the Pacific Islands, and of Australia, I have been led to a simple view as to the origin and affinities of these races. If we draw a line, commencing on the eastern side of the Philippine Islands, thence along the western of Gilolo, through the island of Bouru, and curving round coast the west end of Flores, then bending back round Sandalwood Island to take in Rotti, we shall divide the archipelago into two portions, the races of which have strongly marked distinctive peculiarities. This line will separate the Malayan and Asiatic from the Papuan and Pacific races, and though along the line of junction intermigration and commixture have taken place, yet the division is on the whole almost as well defined and strongly contrasted as are the corresponding zoological divisions of the archipelago into an Indo-Malayan and Austro-Malayan region.

I must briefly explain the reasons that have led me to consider this division of the Oceanic races to be a true and natural one. The Malayan race, as a whole, undoubtedly very closely resembles physically the East Asian populations, from Siam to Mandchouria. I was much struck with this, when in the island of Bali I saw Chinese traders who had adopted the costume of that country, and who could then hardly be distinguished from Malays; and on the other hand I have seen natives of Java who, as far as physiognomy was concerned, would pass very well for Chinese. Then, again, we have the most typical of the Malayan tribes inhabiting a portion of the Asiatic continent itself, together with those great islands which, possessing the same species of large mammalia with the adjacent parts of the continent, have in all probability formed a connected portion of Asia during the human period. The Negritos are no doubt quite a distinct race from the Malay; but yet, as some of them inhabit a portion of the continent, and others the Andaman Islands in the Bay of Bengal, they must be considered to have had, in all probability, an Asiatic rather than a Polynesian origin.

Now, turning to the eastern parts of the archipelago, I find, by comparing my own observations with those of the most trustworthy travellers and missionaries, that a race identical in all its chief features with the Papuan is found in all the islands as far east as the Fidjis; beyond this the brown Polynesian race is spread everywhere over the Pacific. The descriptions of these latter often agree exactly with the characters of the brown indigenes of Gilolo and Ceram; and, moreover, it is to be remarked that the brown and the black Polynesian races have much in common. Their features are almost identical, so that portraits of a New Zealander or Otaheitan will often serve accurately to represent a Papuan, the darker colour and more frizzly hair of the latter being the only differences. I believe, therefore, that the numerous intermediate forms that occur among the countless islands of the Pacific are not merely the result of a mixture of these races, but are truly intermediate or transitional; and that the brown and the black, the Papuan, the natives of Gilolo and Ceram, the Fidjian, the inhabitants of the Sandwich Islands and of New Zealand (and perhaps even of Australia), are all varying forms of one great Oceanic or Polynesian race.

But then arises the question, Whence came this race? Are they from America or from Asia, or did they originate in some one of the islands, and spread from this centre over the whole area they now inhabit? I accept none of these alternatives, for all are beset with innumerable objections. My solution of the difficulty depends chiefly upon the evidence for the existence at a comparatively recent period (geologically speaking) of a Pacific continent, or at least of far more land than now exists there; and also upon the fact of the vast antiquity of the human race. I believe that the Polynesian races are descended from the inhabitants of a land which has now in great part sunk beneath the ocean. The evidence of this subsidence is of two kinds. The first and most important branch of it is to be found in Mr. Darwin's essay on *Coral Reefs*, a work which is one of the most perfect models of inductive reasoning extant, but which, from the very fact of its having at once convinced every scientific reader, led to no discussion, and is therefore far less generally read than it deserves to be. Mr. Darwin proved that the great mass of the coral islands of the Pacific have been produced by subsidence, and the distribution of the islands shows that the subsidence must have occurred over extensive areas, and that every mountain summit, as it successively sank beneath the waters, left an *atoll* to mark its former existence. Almost all the islands of any size and altitude now existing in the Pacific are volcanic, or plainly of volcanic origin, intimating that the elevatory forces have there counteracted the depression which has gone on all around. The distribution of animals among these islands is the second proof of a former union, and it further shows us that the breaking up and subsidence was a comparatively recent operation. There is a very general similarity in their productions over a vast range of latitude and longitude. Even islands so remote as the Sandwich and Society groups have land birds common to the two, as have also the Society and the Fidji Islands. The majority of birds, however, have closely allied representative species in the several islands; but there is reason to believe that the forms and species of birds change more rapidly than those of large mammalia, for while the birds of Java, Sumatra, and Borneo are mostly distinct species from those of the south-eastern parts of Asia, many of the great quadrupeds are identical. There seems every reason, therefore, to believe that man, who existed in Europe along with quadrupeds now extinct, may have also existed at the time when a great continent occupied the tropical portions of the Pacific Ocean.

My argument, therefore, is briefly this. We have a wide region of islands, the inhabitants of which, though not all alike, yet resemble each other in many important points, and are very unlike those of the great continents on each side of them. We have a great deal of evidence to show that this vastly extended archipelago is an area of subsidence, and that at a comparatively recent geological epoch wide spreading lands, perhaps great continents, occupied the site of its now thinly scattered islets. And lastly, we have proof that man has been in existence on the earth fully long enough to have inhabited those lands and continents, and that in our own quarter of the globe he has seen and survived physical changes of equal amount.

Here, then, we have a means of escaping the difficulty of supposing the Oceanic races to have been derived from continents with which they have no communication, and from people with whom they have no affinity. If we accept this view, we need no wonderful migrations in various directions over stormy oceans, we need no new power to introduce rapid changes of physical form and mental disposition; we have only to allow for a certain amount of diversity in the inhabitants of that early land, and the slow but certain effects of the varying physical conditions which have resulted in the present state of its surface, and the diversity of the Polynesian races will be sufficiently accounted for.

In conclusion, I will just allude to the harmony which exists between this division of races and that of the zoological productions of the same region, which I have sketched at the commencement of this paper. The dividing lines do not, it is true, exactly agree; but I think it is a remarkable fact, and something more than a mere coincidence, that they should traverse the same region and approach each other so closely as they do. If, however, I am right in my supposition that the district where the dividing line of the Indo-Malayan and Austro-Malayan regions of zoology can now be drawn was formerly occupied by a much wider sea than at present, and if man existed on the earth at that period, we shall see good reason why the races inhabiting the Asiatic and Pacific areas should now meet and partially intermingle in the vicinity of that dividing line.

It will have been remarked that I have said little about the languages of these races, and have made little use of them in the questions I have been discussing. This is not altogether from want of materials, for I have myself collected vocabularies of about fifty distinct languages of the remoter islands of the Malay Archipelago; but it is because I conceive that the evidence of language is quite subsidiary to that of physical and moral characteristics. It is admitted that language changes much more rapidly than form and character, which must make it of inferior value. Its true importance seems to be in tracing those recent migrations of tribes and intermixture of races of which we have no more direct evidence. Language also gives but uncertain indications, except in particular favourable cases. The illustrious William von Humboldt, from its indications alone, arrived at the conclusion that but one great Malayan race inhabited alike Madagascar, the Malayan, and the Pacific Islands. Mr. Crawfurd, on the other hand, going over the same ground with far more extensive and complete information, determines that the same region is inhabited by several distinct races, speaking some scores of distinct languages. The resemblances which by the first author were taken to show a common origin, are considered by the second only as the result of the wanderings (premeditated or accidental) of the dominant Malays and Javanese, who have left more or less of their languages over this wide region. From a perusal of Mr. Crawfurd's most valuable and interesting *Dissertation on the Affinities of the Malayan Languages*, it seems evident that nearly the same sounds or letters are used for all those on the west of my dividing line, or the true Malayan languages; while the eastern or Polynesian tongues are deficient in some of these sounds, and in their places use others not found to the westward. I have not yet been able to give the necessary time to an examination of my vocabularies, but fully expect that they will show a fundamental difference on this point, and thus serve to confirm the results drawn from a very different kind of evidence.

It will no doubt be said, that all this latter part of my paper is mere speculation. I admit that it is so; but after having had long and familiar intercourse with the races of which I am treating, and having given much consideration to the subject, I trust I may be permitted to speculate a little on their probable origin. Besides, there are speculations and hypotheses of various kinds, and of various degrees of value. There are speculations which are framed to support a foregone conclusion, and which ignore all but the one class of facts which may be deemed favourable. Such are altogether valueless, and deserve all the neglect that they can receive. But when the contriver of a hypothesis has no preconceived opinions to support, when he weighs and sets against each other all the conflicting facts and arguments which bear upon the question, and when his sole object is to discover what supposition will harmonise the greatest number of facts and contradict the fewest, then his speculations deserve some consideration, until they can be overthrown by positive evidence, or until some other hypothesis can be framed which shall, on similar grounds, be better worthy of acceptance.

It is now almost universally acknowledged that true science only begins when hypotheses are framed to express and combine the facts that have been accumulated. And though in this case I am aware that my facts are comparatively scanty, and that they are much wanting in completeness and in precision, yet I have ventured to lay before this society the opinion I have arrived at, that in this case, and I have no doubt in most others, the recent geological changes of the earth's surface have played an important part in determining the present geographical distribution of the races of mankind.

Note Appearing in the Original Work

1. See paper in the Journal of the Royal Geographical Society, 1863, , for fuller details on this subject. [[on]]